# In

# memory

# of

# Albert Einstein's

# miracle year

# 1905

FSC
www.fsc.org
MIX
Papier aus ver-
antwortungsvollen
Quellen
Paper from
responsible sources
FSC® C105338

# Dr. rer. pol. Erik Kolek

# In

# memory

# of

# Albert Einstein's

# miracle year

# 1905

**Chronicles of Business Informatics Physics (CBIP)**

**Volume 4, edition no. 1.0**

**2025**

# Imprint



*Scientific citation*:

Kolek, Erik (2025). In memory of Albert Einstein's miracle year 1905. In: *Chronicles of Business Informatics Physics (CBIP)*. Volume 4, edition no. 1.0. ISBN: 978-3-7693-5395-2.

The Chronicles of Business Informatics Physics (CBIP) consist of cited significant scientific papers and the individually published research articles of the author Dr. rer. pol. Erik Kolek. As an author and an editor of the Chronicles of Business Informatics Physics (CBIP), Dr. rer. pol. Erik Kolek (erster.kontakt@erikkolek.de) is also responsible for editing, translation, typesetting, design (including cover), texts, images and cover picture. Publisher: BoD · Books on Demand GmbH, Überseering 33, 22297 Hamburg, Germany, bod@bod.de. Print: Libri Plureos GmbH, Friedensallee 273, 22763 Hamburg, Germany.
ISBN: 978-3-7693-5395-2

*Bibliographic information of the German National Library*: The German National Library lists this publication in the German National Bibliography; detailed bibliographic data is available on the Internet at dnb.dnb.de. 

Kolek, Erik (2025). In memory of Albert Einstein's miracle year 1905. In: *Chronicles of Business Informatics Physics (CBIP)*. Volume 4, edition no. 1.0. ISBN: 978-3-7693-5395-2.

## Foreword

In this booklet, extensions of the following technical articles by Albert Einstein are described by Erik Kolek:

- On a heuristic point of view concerning the generation and transformation of light. [*Annalen der Physik* 17, 132, (1905a)]
- On the movement of particles suspended in liquids at rest as required by the molecular kinetic theory of heat. [*Annalen der Physik* 17, 549, (1905b)]
- On the electrodynamics of moving bodies. [*Annalen der Physik* 17, 891, (1905c)]
- Does the inertia of a body depend on its energy content? [*Annalen der Physik* 18, 639, (1905d)]
- A new determination of molecular dimensions [*Annalen der Physik* 19, 289, (1906)]

Only Erik Kolek's additions to the specialist articles mentioned are shown. These articles are assumed to be known. These are the well-known articles from Albert Einstein's miracle year 1905. It is now 120 years since these articles appeared in the *Annalen der Physik*. At this time, Albert Einstein was still a patent office employee in Bern. Later in 1921, Albert Einstein received the Nobel Prize for this work, for the photoelectric effect and not for the theory of relativity.

It will also deal with the quantity of light and the speed of light. New equations based on Albert Einstein will be introduced. It will be shown that the speed of light has no limit. The theory of heat is illustrated using temperature. Electrons, i.e. charged particles, cannot reach the speed of light. It also deals with radiation from neutron stars and black holes. Planar gravitational waves are discussed in this context. A new determination of molecular dimensions can be calculated more precisely. The extensions of the five *Annalen der Physik* articles shown here only correspond to the opinions of Erik Kolek. They are possible starting points which, taken together, provide useful research results. Knowledge of the mathematics of theoretical physics is required to understand the content.

Kolek, Erik (2025). In memory of Albert Einstein's miracle year 1905. In: *Chronicles of Business Informatics Physics (CBIP)*. Volume 4, edition no. 1.0. ISBN: 978-3-7693-5395-2.

The reader is advised to read the extensions in the order given. Long content sentences can be read with a little perseverance and diligence. However, the technical articles must be read first, otherwise no understanding of the subject matter can be gained. The latter articles are not quoted or printed here in full for copyright reasons. All of Erik Kolek's extensions are theories of theoretical physics. These texts were modeled using approaches from business informatics. These model representations are presented below.

Schwäbisch Hall, May 09, 2025.

Kolek, Erik (2025). In memory of Albert Einstein's miracle year 1905. In: *Chronicles of Business Informatics Physics (CBIP)*. Volume 4, edition no. 1.0. ISBN: 978-3-7693-5395-2.

Erik Kolek

Doctor of Economic Science (Dr. rer. pol.)

Diplom-Betriebswirt (FH), M.A., M.Sc.

Kolek, Erik (2025). In memory of Albert Einstein's miracle year 1905. In: *Chronicles of Business Informatics Physics (CBIP)*. Volume 4, edition no. 1.0. ISBN: 978-3-7693-5395-2.

## Table of contents

*Scientific citation*:

Kolek, Erik (2025). On a heuristic point of view concerning the generation and transformation of light – An extension. In: *In memory of Albert Einstein's miracle year 1905*. Chronicles of Business Informatics Physics (CBIP). Volume 4, edition no. 1.0.

## On a heuristic point of view concerning the generation and transformation of light – An extension; by E. Kolek.

First, we transform the latter relationship into L and obtain the amount of light absorbed L:

$$L = jR\beta v$$

The latter equation applies to the ionization of gases by ultraviolet light. This statement can be confirmed by experiment. For $R\beta v$ we now insert $IIE + P'$ and obtain:

$$L = (IIE + P')j$$

This is the function of the frequency of the exciting light described in Cartesian coordinates. The light path here represents a straight line. E is the charge of a gram equivalent of a monovalent ion and P' is the potential of this quantity of positive electricity with respect to the light. The following can be said to determine the speed of light:

$$L = \sqrt[3]{\frac{R}{N}\frac{8\pi v^2}{\varrho_v}T}$$

or

$$L = \sqrt[3]{\hat{E}\frac{8\pi v^2}{\varrho_v}}$$

Kolek, Erik (2025). In memory of Albert Einstein's miracle year 1905. In: *Chronicles of Business Informatics Physics (CBIP)*. Volume 4, edition no. 1.0. ISBN: 978-3-7693-5395-2.

We therefore obtain the following equation for the limit of the speed of light:

$$\int_0^\infty L^3\,dv = \hat{E}\,\frac{8\pi}{\varrho_v}\int_0^\infty v^2\,dv = \infty.$$

It is clear that there can be no limit to the speed of light L, because it is $\infty$ infinite. $\hat{E}$ is a dependency here, because if $\hat{E}$ increases, the speed of light L also increases. You can also write for L:

$$L = \sqrt[3]{\frac{\beta}{\alpha}\,\frac{8\pi R}{N}}$$

We come to the conclusion that the speed of light depends on the gravity of the atoms. The heavier (lighter) the atoms are, the slower (faster) is the speed of light and vice versa.

Schwäbisch Hall, May 07, 2025.


## Reference

Einstein, A. (1905a). Über einen die Erzeugung und Verwandlung des Lichtes betreffenden heuristischen Gesichtspunkt. *Annalen der Physik* 17, 132.

*Scientific citation:*

Kolek, Erik (2025). On the movement of particles suspended in liquids at rest as required by the molecular kinetic theory of heat – An extension. In: *In memory of Albert Einstein's miracle year 1905.* Chronicles of Business Informatics Physics (CBIP). Volume 4, edition no. 1.0.

**On the movement of particles suspended in liquids at rest as required by the molecular kinetic theory of heat – An extension; by E. Kolek.**

We use the relationship found to determine the temperature T, as this is important for establishing the theory of heat:

$$T = \frac{\lambda_x^2 \, 3\pi k P N}{tR}$$

The time t is therefore important when it comes to determining the temperature T. This is a new method for determining the true temperature T of atoms. The temperature T can also be determined by the diffusion coefficient.

$$T = - DN6\pi kP$$

The absolute temperature T is negative here because the diffusion coefficient of the suspended substance is negative. The temperature T can also be determined via the osmotic pressure p.

$$T = \frac{pN}{Rv}$$

The osmotic pressure p has an influence on the temperature T. The temperature T is therefore dependent on the total number of particles N and their velocity v as dissolved material in liquids. The following can be determined using the free energy:

$$T = \frac{-FN}{lgBR}$$

Kolek, Erik (2025). In memory of Albert Einstein's miracle year 1905. In: *Chronicles of Business Informatics Physics (CBIP)*. Volume 4, edition no. 1.0. ISBN: 978-3-7693-5395-2.

N is the number of actual molecules contained in one gram of molecule. The free energy F of these molecules N therefore determines our theory of heat. It can also be formulated in general terms:

$$T = \frac{pV^*}{Rz}$$

The latter corresponds to the classical theory of thermodynamics, with gravity being neglected. The question raised here with regard to the theory of heat has been examined in more detail, but it remains open to decide this important question.

Schwäbisch Hall, May 07, 2025.


## Reference

Einstein, A. (1905b). Über die von der molekularkinetischen Theorie der Wärme geforderte Bewegung von in ruhenden Flüssigkeiten suspendierten Teilchen. *Annalen der Physik* 17, 549.

Kolek, Erik (2025). In memory of Albert Einstein's miracle year 1905. In: *Chronicles of Business Informatics Physics (CBIP)*. Volume 4, edition no. 1.0. ISBN: 978-3-7693-5395-2.

*Scientific citation*:

Kolek, Erik (2025). On the electrodynamics of moving bodies – An extension. In: *In memory of Albert Einstein's miracle year 1905*. Chronicles of Business Informatics Physics (CBIP). Volume 4, edition no. 1.0.

**On the electrodynamics of moving bodies – An extension; by E. Kolek.**

The following statement is of particular interest: "Superluminal velocities have [...] no possibility of existence" (Einstein, 1905c, p. 920). This hypothesis can either be accepted or rejected. As an example, an electrically charged particle, i.e. an electron, will be used here to determine the speed. $W = v = V$ applies.

$$\frac{A_m}{A_e} = \frac{v}{V} = 1 \ (because \ \frac{V}{V}) = A_m = A_e$$

The magnetic deflectability $A_m$ is equal to the electric deflectability $A_e$ in the limiting case of superluminal velocity. This relationship is accessible to experiment. Oscillating electric and magnetic fields can be used for this purpose. The electric force $Y$ is then equal to the magnetic force N, since $Y = N \times v/V = N$. (A) applies to the motion of the electron:

$$\frac{d^2x}{dt^2} = \frac{\epsilon}{\mu}\frac{1}{\beta^3}X,$$

$$\frac{d^2y}{dt^2} = \frac{\epsilon}{\mu}\frac{1}{\beta}(Y - N),$$

$$\frac{d^2z}{dt^2} = \frac{\epsilon}{\mu}\frac{1}{\beta}(Z + M).$$

The same applies to the potential difference:

$$P = \frac{\mu}{\epsilon}V^2\left\{\frac{1}{\sqrt{1 - (1)^2}} - 1\right\}$$

Kolek, Erik (2025). In memory of Albert Einstein's miracle year 1905. In: *Chronicles of Business Informatics Physics (CBIP)*. Volume 4, edition no. 1.0. ISBN: 978-3-7693-5395-2.

and for the radius of curvature R =

$$V^2 \frac{\mu}{\in} \times \frac{1}{\sqrt{1-(1)^2}} \times \frac{1}{N}$$

For the kinetic energy W you get:

$$W = \mu V^2 \left( \frac{1}{\sqrt{1-(1)^2}} - 1 \right)$$

W therefore remains infinitely large if W = v = V. A slowly accelerated electron can therefore not reach superluminal velocity. Especially not because it is a ponderable mass whose masses are determined as follows:

$$Longitudinal\ mass = \frac{\mu}{(\sqrt{1-(1)^2})^3}$$

$$Transverse\ mass = \frac{\mu}{1-(1)^2}$$

Schwäbisch Hall, May 8, 2025.


## Reference

Einstein, A. (1905c). Zur Elektrodynamik bewegter Körper. *Annalen der Physik* 17, 891.

Kolek, Erik (2025). In memory of Albert Einstein's miracle year 1905. In: *Chronicles of Business Informatics Physics (CBIP)*. Volume 4, edition no. 1.0. ISBN: 978-3-7693-5395-2.

*Scientific citation*:

Kolek, Erik (2025). Does the inertia of a body depend on its energy content? – An extension. In: *In memory of Albert Einstein's miracle year 1905*. Chronicles of Business Informatics Physics (CBIP). Volume 4, edition no. 1.0.

## Does the inertia of a body depend on its energy content? – An extension; by E. Kolek.

Radiation can transfer inertia between emitting and absorbing bodies. "If a body emits the energy L in the form of radiation, its mass decreases by $L/V^2$" (Einstein, 1905d, p. 641). If the energy changes by L, then the mass changes by L. The mass of a body is therefore a measure of its energy content. The latter applies equally to stars and black holes.

$$K_1 = K_0 - \frac{L}{V^2}\frac{v^2}{2} = K_0 - \frac{L}{2}$$

It applies to the speed of light v = V, especially for superluminal speeds. Light radiation causes the kinetic energy of a body to decrease. If a body absorbs the energy L as radiation, its mass increases by $L/V^2$. Increased absorption of radiation causes an increase in mass and vice versa.

$$H_1 + \frac{L}{\sqrt{1-(\frac{v}{V})^2}} - \left[E_1 + \left(\frac{L}{2}+\frac{L}{2}\right)\right] - \left[K_1 + L\left(\frac{1}{\sqrt{1-(\frac{v}{V})^2}} - 1\right)\right] = C$$

$$H_1 - E_1 - L - K_1 + 1 = C$$

$$L = H_1 - E_1 - K_1 + 1 - C$$

A black hole can absorb energy L in the form of radiation and emit it at the same time, resulting in the emission of light rays. Its mass will increase and decrease by $L/V^2$. A star, on the other hand, will emit energy L and lose mass by $L/V^2$.

Kolek, Erik (2025). In memory of Albert Einstein's miracle year 1905. In: *Chronicles of Business Informatics Physics (CBIP)*. Volume 4, edition no. 1.0. ISBN: 978-3-7693-5395-2.

The earth absorbs radiation in the form of light energy L, thereby increasing its mass by $L/V^2$. The light is therefore part of the gravitational effect. The same can also be assumed for the Earth's moon and other celestial bodies.

Light waves can be equivalent to gravitational waves. Both types of waves can be seen. The following equation could then apply to gravity:

$$g^* = g\frac{1 - \frac{v}{V}\cos\varphi}{\sqrt{1 - (\frac{v}{V})^2}}$$

V stands for the speed of light. If a body emits gravitational waves, then the following applies:

$$E_0 = E_1 + [\frac{G}{2} + \frac{G}{2}]$$

and

$$H_0 = H_1 + \frac{G}{\sqrt{1 - (\frac{v}{V})^2}}$$

The following is still valid:

$$(H_0 - E_0) - (H_1 - E_1) = G\{\frac{1}{\sqrt{1 - (\frac{v}{V})^2}} - 1\}$$

Since gravitational wave radiation is constant, the following applies:

$$K_0 - K_1 = G\{\frac{1}{\sqrt{1 - (\frac{v}{V})^2}} - 1\}$$

and

Kolek, Erik (2025). In memory of Albert Einstein's miracle year 1905. In: *Chronicles of Business Informatics Physics (CBIP)*. Volume 4, edition no. 1.0. ISBN: 978-3-7693-5395-2.

$$K_0 - K_1 = \frac{G}{V^2} \frac{v^2}{2}$$

If a body emits gravity G in the form of waves, its mass is reduced by $G/V^2$. If the gravity changes, the mass changes. The latter could apply to neutron stars in particular, as they emit gravitational waves in pulses. It cannot be ruled out that this relationship also applies to black holes as well as to stars.

Schwäbisch Hall, May 09, 2025.


## Reference

Einstein, A. (1905d). Ist die Trägheit eines Körpers von seinem Energieinhalt abhängig?. *Annalen der Physik* 18, 639.

Kolek, Erik (2025). In memory of Albert Einstein's miracle year 1905. In: *Chronicles of Business Informatics Physics (CBIP)*. Volume 4, edition no. 1.0. ISBN: 978-3-7693-5395-2.

*Scientific citation*:

Kolek, Erik (2025). A new determination of molecular dimensions – An extension. In: *In memory of Albert Einstein's miracle year 1905.* Chronicles of Business Informatics Physics (CBIP). Volume 4, edition no. 1.0.

**A new determination of molecular dimensions – An extension; by E. Kolek.**

Following Albert Einstein (1906, p. 304), the following applies: "The osmotic force acting on the unit of mass" $\frac{1}{\varrho}\frac{\partial p}{\partial x}$ "can be caused by the force (acting on the individual dissolved molecules)" $P_x$ "the balance can be achieved if"

$$+\frac{1}{\varrho}\frac{\partial p}{\partial x} + P_x = 0$$

Albert Einstein achieved in his § 3:

$$P^3 = \frac{\frac{K^*}{K} - 1}{n\frac{1}{3}\pi}$$

$K^*$ is the coefficient of friction of the solution. K is that of the solvent. Substituting this into the following equation gives:

$$NP^3 = \frac{3}{4\pi}\frac{m}{\varrho}\left(\frac{K^*}{K} - 1\right)$$

$$N\left(\frac{\frac{K^*}{K} - 1}{n\frac{1}{3}\pi}\right) = \frac{3}{4\pi}\frac{m}{\varrho}\left(\frac{K^*}{K} - 1\right)$$

$$N\left(\frac{K^*}{K} - 1\right) = n\frac{1 \times 3\pi}{3 \times 4\pi}\frac{m}{\varrho}\left(\frac{K^*}{K} - 1\right)$$

Kolek, Erik (2025). In memory of Albert Einstein's miracle year 1905. In: *Chronicles of Business Informatics Physics (CBIP)*. Volume 4, edition no. 1.0. ISBN: 978-3-7693-5395-2.

$$N = n\frac{3}{12}\frac{m}{\varrho}$$

$$N = \frac{n}{4}\frac{m}{\varrho}$$

n is the number of dissolved molecules per unit volume. $\varrho$ is the mass of the solute in the unit volume. m is the molecular weight. Now P and N can be calculated individually.

$$N\varrho = nm = N = \frac{nm}{\varrho}$$

N can therefore be determined more precisely. N can be used in the equation for the diffusion coefficient. The following applies:

$$NP = \frac{RT}{6\pi k}\frac{1}{D}$$

$$P\left(\frac{nm}{4\varrho}\right) = \frac{RT}{6\pi k}\frac{1}{D} = (\frac{RT}{6\pi k}\frac{1}{D}\frac{4\varrho}{1})/nm$$

$$P = \frac{\frac{4RT\varrho}{6\pi kD}}{nm}$$

$$P = \frac{\frac{2RT\varrho}{3\pi kD}}{nm}$$

P is the hydrodynamically effective molecular radius. N is the number of actual molecules. The latter can now be determined more precisely.

Schwäbisch Hall, May 09, 2025.

Kolek, Erik (2025). In memory of Albert Einstein's miracle year 1905. In: *Chronicles of Business Informatics Physics (CBIP)*. Volume 4, edition no. 1.0. ISBN: 978-3-7693-5395-2.

## Reference

Einstein, A. (1906). Eine neue Bestimmung der Moleküldimensionen. *Annalen der Physik* 19, 289.

Kolek, Erik (2025). In memory of Albert Einstein's miracle year 1905. In: *Chronicles of Business Informatics Physics (CBIP)*. Volume 4, edition no. 1.0. ISBN: 978-3-7693-5395-2.

Notes

Kolek, Erik (2025). In memory of Albert Einstein's miracle year 1905. In: *Chronicles of Business Informatics Physics (CBIP)*. Volume 4, edition no. 1.0. ISBN: 978-3-7693-5395-2.

Notes